Radio Ministry Handbook

Charlie Edwards, PhD

Certified Radio & TV Broadcast Engineer

Certified Audio Engineer

RADIO MINISTRY HANDBOOK
Published by Edwards Ministries, Inc., Chattanooga, Tennessee.

Copyright © 2013 Charlie Edwards
All rights reserved.
ISBN: 978-0-9914146-2-8

Copyright © 1983 Charles K. Edwards
First printing: 1983, Calvary Baptist Publications, Lakeland, Florida

Other books by Charlie Edwards:
CARIBBEAN SENTINEL (novel),
BIBLE WHITE PAPERS, Edwards Ministries, Inc.
CHURCH PA SYSTEM HANDBOOK, Edwards Ministries, Inc.
Published in the United States of America

Cover design by author. The tower drawing on the cover was part of the original cover on the first edition from 1983. All drawings were done by author and are also from the first edition.
Photo on back is author, circa 1981.

DEDICATION

This book is dedicated to Pastors who still see the value of the local radio station as a means to reach the unsaved and un-churched multitudes in our country and around the world.

CONTENTS

	Acknowledgments	i
	Introduction	1
1	The Purpose of the Broadcast	3
2	Choosing a Station – Some Considerations	5
3	Counting the Cost	12
4	The Program Itself	15
5	Equipment Considerations	21
6	Phone Lines	35
7	Glossary	39
8	Audio Equipment Resources	57
9	A word about the author	58

ACKNOWLEDGMENTS

I am grateful to the many radio and TV coworkers who contributed to the information compiled in this small project. I would also like to thank my long-time radio friend, Tommy Sneed for suggesting an updated version of this booklet. Finally, I wish to express my thanks to the folks at Sweetwater.com for granting permission to use some of their definitions in our glossary section. All quotes from Scripture come from the Authorized Version.

Introduction

This booklet was first published in 1983 for the benefit of Gospel Preachers who felt led to use the local radio station as an outreach ministry. At that time I had worked in several radio stations and helped several preachers get started in this wonderful ministry.

Over the course of the past thirty one years, radio has changed considerably. The FCC has changed many former requirements for operation of stations. Ownership laws have changed as well as the equipment and its operation. Most of the people who worked in radio when this booklet was first published are no longer there. There is a whole new generation of radio station owners and operators today. Countless changes have taken place. However, there are a few things that have not changed:

1. The Gospel of Jesus Christ: The good news that Jesus Christ saves souls today just like He did in years past

gives credence to a more up to date version of Radio Ministry Handbook.
2. The Great Commission still commands us to **"Go ye into all the world, and preach the gospel to every creature"** Mark 16:15. The local church is still in the business of reaching out to our community with Christ's salvation message.
3. There are still Christian radio stations that need to partner with local churches and air their programs.
4. Finally, there are still preachers who are interested in this outreach ministry that need some technical assistance.

So as you can see many changes have taken place, but there are still some things that haven't. Therefore it is hoped that the information contained in this boolet will be of assistance to preachers who are not familiar with the radio media. It is also hoped that even those seasoned veterans of radio will find something useful which they possibly can apply to their radio ministries.

CHAPTER 1

The Purpose of the Broadcast

The real purpose of your radio ministry needs to be determined from the start. What is it that you want to accomplish? What audience will you target? What is your goal? Do you want to promote your church and put it before the public? Do you want to entertain Christians with music and devotional material? Is your program to be used as a bulletin board for your church members? Do you want to edify the saints with encouraging thoughts and songs? Do you want to evangelize your town using the radio media? Do you simply want to hear yourself on the radio?

I know a pastor who has a fifteen minute daily program during morning drive time who uses this time to encourage his people with a few Gospel songs, announcements, and a very brief devotional thought. While this pastor directs his message and thoughts to his own church members, he needs to also realize there could possibly be thousands of people who aren't his members (especially during drive time) that could use encouraging also.

Another example is another local pastor who does not play any music on his fifteen minute program. Instead he gets right into a rigid evangelistic salvation message on every program. One needs to keep in mind the audience that listens to the average Christian station that airs local preachers. Most of them are already "saved and churched."

For the larger syndicated programs that go beyond the local community radio station, the goals are different. They are not really looking to bring in new church members. They look more to edifying the Body of Christ and/or win the lost and seek monetary support for the local program costs.

Every pastor is different and will have a different style of program. Obviously, this is a good thing. It's good for the Radio Preacher to keep in mind who his audience is and what he wants to accomplish with his radio program.

CHAPTER 2

Choosing a Station – Some Considerations

In most radio markets there is a variety of radio formats. If your town is of any size at all there should be many types of radio stations. There is a good chance that there will be one or more (depending on where you are in the country) Christian radio stations. There are all types of Christian radio stations. You must choose the one you feel will yield the best results.

Sometimes it might be better to get on a secular station. Most will not allow you any time with the exception of Sunday. In that case a thirty or sixty minute program would be ideal. The only reason they allow you time is to meet their requirement for the Federal communication Commission. They need your program to meet that need.

Sometimes this information can be of help in obtaining a good rate and time.

Here are some items to consider in deciding on a station:

Is the station AM or FM? Both have their advantages and disadvantages. AM stands for amplitude modulation. Without getting side-tracked and giving an explanation too technical for the average preacher to understand, let me simply give some pros and cons.

Some AM stations do not have 24-hour authorization and must go off the air at sundown. They sign on at sunrise. Of course during summer months the broadcast day is longer. Many stations will have a pre-sunrise authorization at a reduced power. Sometimes this early morning day part can be utilized at a reduced rate. I would certainly not want to go on a station with reduced power and not have a reduced rate. This time is generally between 6 AM and local sunrise.

A large number of AM stations are required to reduce power when the sun goes down and resume normal power when the sun comes up. This is due to the ground wave propagation and sky wave propagation. More details on this are available on the FCC website: http://www.fcc.gov/encyclopedia/why-am-radio-stations-must-reduce-power-change-operations-or-cease-broadcasting-night

The maximum power allocation for an AM station in the standard broadcast band is 50,000 watts. Certainly it is advantageous to obtain time on a station with this much power. Generally the 50 kilowatt stations have to go off the air or reduce power at sundown. Many reduce power and go to a directional antenna at this time. There are still some that operate into a directional antenna all of the time. Only a few of these stations operate non-directional twenty-four hours a day. These are known as Clear Channel Stations. You can hear them mostly at night. This is due to the action of the sun on the ionosphere (see figure 1 and 2). It goes without saying that the clear channel radio station, if the time were available, would be profitable to be on.

The frequency of an AM station has much to do with its coverage area. A station operating at the lower end of the band (from 535 KHz to 1070 KHz) would be, for example, more advantageous than a station at the upper end of the band (from 1070 to 1605 KHz) (see figure 3). Essentially, the higher in frequency you go, the less distance you are able to transmit with the same amount of power. For example, a 5000 watt station operating on 540 KHz will have a greater coverage area than a 5000 watt station operating on 1530 KHz. This is especially true at night. So, for the AM station, your lower frequencies are going to give you better coverage than the higher ones. Certainly, you would not want to choose a 500 watt station on 600 KHz over a 5000 watt station on 1450 KHz. If the power were the same for both stations you would be better off with the station on 600 KHz (with respect to coverage area).

Another consideration is frequency response. The frequency response for the human ear is from 20 Hz. to 20,000 Hz. This will vary depending on the individual.

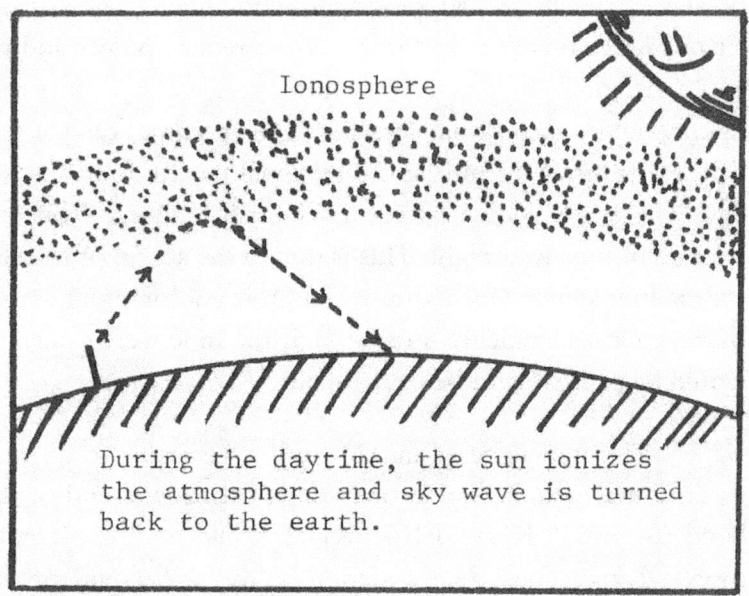

Figures 1 and 2

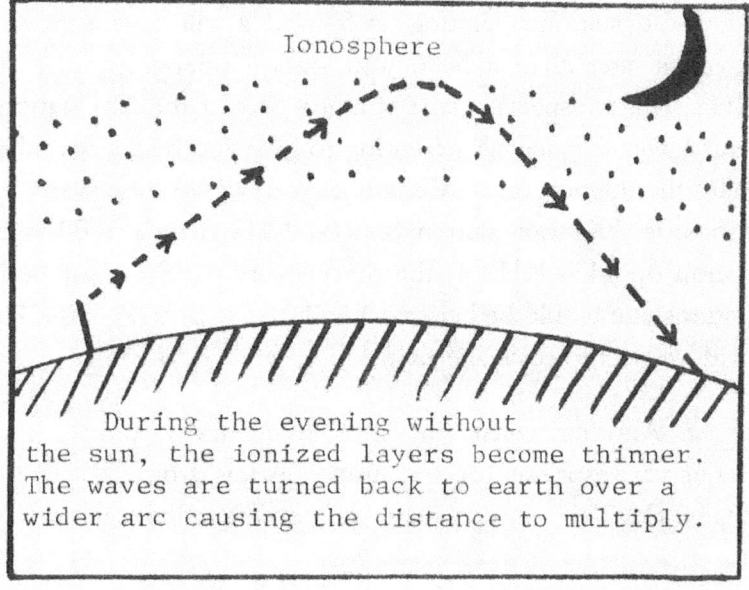

Normally the high end will drop off with age. This also depends on "noise abuse." For example a factory worker (in a noisy environment) that doesn't wear hearing protection will most likely have hearing problems later in life.

The frequency response of an AM station is 50 – 7,500 Hz. The frequency response of an FM station is 50 – 15,000 Hz (see figure 4). As you can see, the FM frequency response is twice that of the AM. This is the reason FM sounds more pleasing to the ear than AM. It has a better frequency response. If you want to use recorded music on your program and want that music to sound optimum, then you should try to get on an FM station. If you want the emphasis to be upon the spoken message or voice frequencies (200 – 3000 Hz.), then either AM or FM will work quite well. This is not to say that AM will not work well for music, but that FM will work better.

FM stands for Frequency Modulation. This is completely different than Amplitude Modulation. Again, without getting technical, FM differs from AM in that the carrier frequency is modulated by varying the frequency instead of the amplitude. What does that mean? Well, for one thing, the interference level is less with FM than AM. When you listen to an FM station in a thunder storm it will not be disturbed near as much as an AM station would be. Another factor is the surrounding environment. A mountain, for example, can affect an FM signal more than an AM signal. If you are located in a mountainous area perhaps an AM station would give better coverage. Again, have you ever driven through a tunnel or gone under a bridge and lost the AM signal you were listening to? Have you noticed the FM signal does not disappear (unless the tunnel is far beneath a mountain; even then reception will be retained much longer than AM). This is due to the wave length involved. Briefly, the wave length for an FM station is approximately 18 to 24 inches long. The wave length of an AM wave is several city blocks long. The average tunnel, large enough to drive a truck through would offer plenty of room for an FM signal to pass through, but not an AM signal.

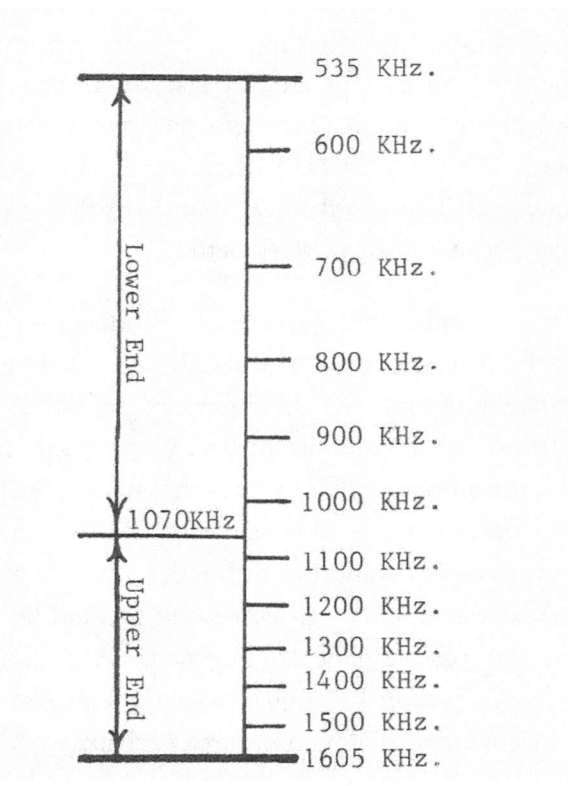

Figure 3

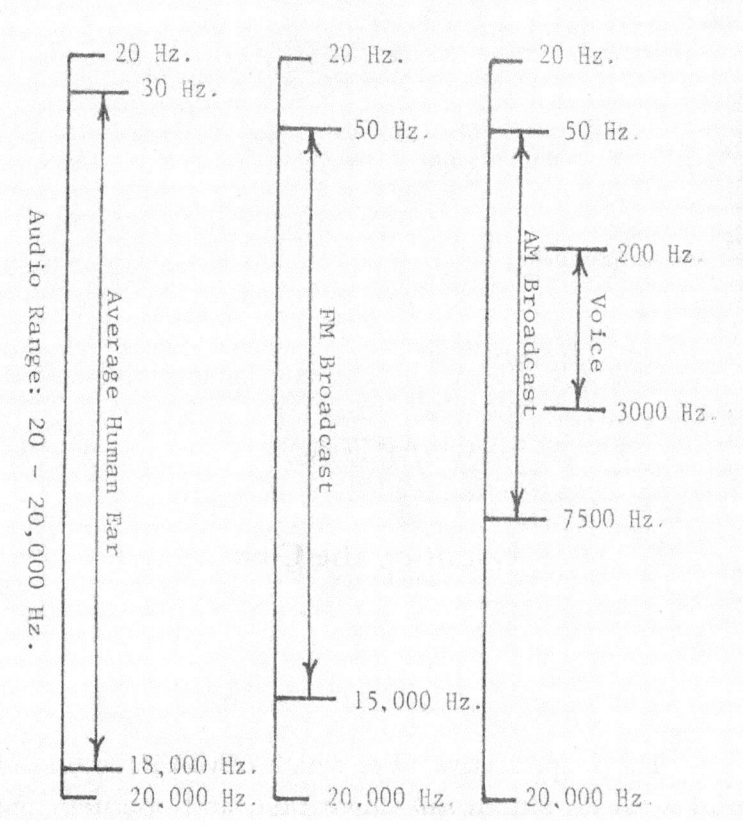

Figure 4

The maximum power allocation for FM stations is 100,000 watts. An FM station does not have to cut back power at sundown. Basically, the coverage will be the same for both day and night for FM. When enquiring about a broadcast at a station, find out about both night time and day time coverage.

CHAPTER 3

Counting the Cost

The Scripture says, **"For which of you intending to build a tower, sitteth not down first, and counteth the cost, whether he have sufficient to finish it?"** (Luke 14:28). I have seen many preachers start a radio broadcast and quit after four to six weeks. The reason? They did not realize the time element involved, the financial element, etc. It takes much just to keep up with a five-minute daily program. It takes more to do a fifteen-minute broadcast and even more for a thirty-minute daily. Count the cost before you start.

Finances

It would be wise to decide ahead of time where the

finances are going to come from for the program. One idea is to use the Wednesday night offering to pay for the air time. Another method is to add it to the church budget, perhaps the missionary budget. Sometimes individuals will volunteer to pay for a radio broadcast. A church that is always up to date on its bill at the radio station is an excellent testimony to an unsaved station manager. There are many preachers preaching on the air over radio stations across this country with overdue balances on their accounts. This is shameful. Don't let this happen to you.

In addition to air time, costs should be figured for equipment and supplies. Quality equipment costs money. It would be better to put off going on the air and using that money to invest in decent radio gear. As the money comes in for the radio ministry, purchase the equipment you need to do the job properly, even if it means putting off the start date a year.

Not long ago I was speaking with a Baptist preacher who wanted to start on the air with a fifteen minute daily program. I encouraged him to start immediately with a special radio offering from his people. In only a few weeks he could be able to purchase the equipment needed and not have to struggle with inferior equipment. This plan has worked for many preachers who are on the air now putting out quality programs with quality equipment.

Time

Another cost to count is time. One does not realize the time it takes to prepare a fifteen minute daily program until one does it for a period of time. If yours is a live program, you must figure in driving time to the radio station on top of other preparation time. Multiply this by five days per week and you have a large chunk taken out of your schedule each week for the radio ministry. One way to get around this is the use of phone lines. This will be covered in more detail in a later chapter. If you are planning on using recorded broadcasts, you must realize that it takes time. The first couple of times a novice radio preacher errs during his recorded broadcast, he may want to go back and begin again to correct the mistake. After about four repeats of this procedure he will learn the value of time. Count the cost in time.

Postage

Many radio preachers record an entire week's programs at one time and deliver them to the station once per week. If mailing the tapes or CDs to the station should be required, then the postage costs should be figured into the radio budget. Money can be saved by mailing ahead of time and utilizing the slower postage rates. Fourth class and media rate information can be obtained from your local post office. Recording programs ahead of time assures your voice on your program should you have to be out of town.

It's better to count the cost on each of these items before you go on the air and know what to expect. Consistency is a great way to build a testimony for your church in your town. Once you start your broadcast, you should plan on continuing until Jesus comes.

CHAPTER 4

The Program Itself

After you have decided on a purpose for your broadcast; after you have chosen your station and counted the cost; what's left? How about a title for the program? Seriously, much thought should go into the choice of a title. The title of your program will be what many people will think of when they hear your voice on the air. I worked at a station where the management was putting five-minute daily programs on the air utilizing some of the local preachers. The first one called his program, "Moment of Praise." The next preacher dubbed his "One Moment Please." Finally the third one labeled his program, "Wait a Minute!" Somebody got hung up on minutes. Sure it's only five minutes a day, but use some imagination! Try not to duplicate titles of other programs on in the same radio market.

The title should be simple and uncomplicated. It is a good idea to use the talent of a professional announcer to produce an intro and outro for your program. In most cases someone at the station can do this and is included with the program. Many times a church member who "ought to be in radio" will record this and will over-do it with an exaggerated "radio voice." Use a professional voice. The brief announcement should contain all the vital information; including who, what, where, etc. After your program goes off, the listener should know who you are and where and how they can get in touch with you or your church.

The time of day your program is on has much to do with the size of your listening audience. Of course, you will want the time when most people will be listening. Unfortunately, this is not always available. The station management also knows when the audience is listening the most. The station manager might prefer that time to be allotted for programming that is more conducive to generating revenue such as music and spot sales. At any rate, the two most listened to times of the broadcast day are from 6 – 10 AM and from 4 – 7 PM. These times are known as morning and afternoon drive time. The reason for this is simple. There are more people in their cars listening to the radio, coming home or going to work, during this time than any other. Normally, during these times or day-parts, the station will require the most money per second or minute. If your church has the money to get on during this time and it is available, get it!

In a previous chapter, we mentioned the difference in coverage of the day-night operation of AM stations. In many cases during the evening hours an AM station covers a larger area than it does during the day. This all depends upon the power and antenna system of the station, but would be worth checking into. A church broadcast wants as much coverage as possible. This could be one avenue to obtain more coverage. In choosing an AM station, one must also realize that the signal output might decrease with the night time power. Some stations might have 1000 watts during the day and 250 or 500 at night. During the summer this change will take effect at about 9 PM, while in the winter the sun sets at about 5:30 PM. This should be taken into consideration as well. The power change times fluctuate each month for AM stations, moving gradually toward each extreme. In other words, your 7 PM program, during the summer month (before sundown) will get more coverage than at the same time six months later (after sundown).

Many times small market stations will throw in a free Sunday time slot as part of a daily "package." In plainer words, if you are on for fifteen or thirty minutes daily and pay a good rate, then the station management might allow you a free thirty or sixty minute program on Sunday. This is just an example. Each station has its own style of wheeling and dealing with the package deals. Simply by inquiring you might be able to get a free time (if you are thinking in terms of a daily program).

There is also another service a station can offer its clients. This is promotional announcements for the program itself. Perhaps this could even be a promotion for the church. This is a thirty or sixty second spot announcement telling what time the program comes on; when, where, whom, etc.

This would be aired at some time of the day other than the time of the program. For example, if you were on daily, you might receive a free thirty second spot during drive time to promote your program. It can't hurt to ask about it. All they can say is, "Sure, no problem!"

Something to be concerned about also is adjacencies. Who is on before you and who comes on after you? No preacher wants to have someone from another denomination come on after his program is finished and contradict everything he just said. Unfortunately, this happens. It would pay to find out who is on after you.

When you inquire about a program at a station you will want to know the advantages and disadvantages of program length. If you are on daily with a five minute program, you will not be able to cover very much scripture. Many use a five minute slot simply to put their church before the public and give the audience something to lure them into the church on Sunday. A fifteen minute daily gives the preacher a little more to work with. This allows enough time for church promotion, music, and a devotional. I have listed below some simple examples of radio formats according to time allotment.

Typical format for a five minute daily:

Intro	:30
Church promotion	4:30
Outro	:30
Total	5:00

Typical format for a fifteen minute daily:

Intro	:30
Greeting	:60
Song	3:30
Announcements	3:00
Prayer	2:00
Message	4:30
Outro	:30
Total	15:00

Typical format for a thirty minute daily:

Intro	:30
Greeting	:60
Song	3:30
Announcements	3:00
Song	3:30
Prayer	3:00
Message	15:00
Outro	:30
Total	30:00

Typical format for sixty minute weekly:

Into	:30
Excerpt from last Sunday's service recorded (music and message in full)	57:30

Outro	:30

Note: A longer outro could be used on full length messages to offer a recording or transcription of the message, offer books, invite listeners to church etc.

Some churches have a live broadcast on Sunday morning or evening. This is fine as long as your services are not subject to the broadcast instead of just the opposite. One thing to remember in this case is the monthly payment on the phone line to the radio station. By recording the service and playing it back next week or later the same day, one finds some advantages:

- No phone line costs
- Not being hampered by pressure to start or end on time or censorship.
- No "technical difficulties"
- Your choice of messages. If your program is on Sunday morning and you bomb the message in the morning, but the evening message was anointed with God's power; you have the choice of which to put on the air.
- Option to "cut out" and edit portions of the message or service not wanted on the air.
- Complete your own list.

CHAPTER 5

Equipment Considerations

The quality of your audio equipment will factor into the success of your radio ministry. Professional equipment does make a difference. Many times the first short-cut in the budget is the sound equipment. In days past this was common. I have found that more and more churches are seeing the need for better audio gear and are doing something about it. If you are planning on a radio ministry or even in the planning stage of a new church PA system, you will be better off in the long run if you spend more money up front.

You have probably heard a Gospel broadcast in which the quality was nonexistent. This is a common occurrence on "dollar-a-holler" Gospel radio stations. I'm talking about low budget programs that the signal to noise ratio is so low that you can barely hear the preacher over the noise. The program that you produce on your equipment is the finished product that will go before the public. If it has good quality it will be received more readily than poor quality. All one need do to prove this is look at the more successful syndicated radio programs and their methods. These are examples of successful radio ministries which have discovered the importance of quality audio equipment: Decision Minute, Grace to You, Hour of Decision, Insight for Living, Love Worth Finding, Southwest Radio Church Ministries, Thru the Bible, and so many more.

I mentioned some things concerning the financing of the program and the equipment needed in an earlier chapter. I suggested using the Wednesday evening offering. Some preachers put a box in the church foyer and ask members to contribute by simply putting money for the radio ministry into the box. Some add it to the mission budget; others add the expense to the regular church budget. One method I did not mention was listener support. This is the most used method among the larger syndicated programs. Listeners in each station's coverage area support that particular area. Many of these use books, CDs, DVDs, etc. to make it "easier" for the listener to give his money. Use something which can be obtained inexpensively, in quantity, and offer it for a gift to the listener in your radio ministry. If you are on a non-commercial station you must ask for an offering in exchange for the free gift. This will pay for air time for that particular station. Selling items for a specific price may violate the non-commercial station's license. Even commercial stations might want you to list the amount of time that you "sell" on your program for their program log records. It is best to stay with the "gift for the ministry" approach. The reason I have included this financial information in the equipment chapter is that it may be used for equipment updating as well as payment for air time.

Microphones

The correct choice of a microphone to record a broadcast can make a big difference in the on-air sound. It is best to stay with low impedance mikes. Again, without getting too technical, there are two types of microphones (as far as impedance is concerned); Low and High Impedance. Low impedance is generally used in broadcast and recording industries as well as public address systems. Regardless of whether you will be plugging your mike directly into the tape recorder, into a mike mixer, or an individual mike preamp (to drive a phone line), it is best to stick with Low Impedance microphones.

Another characteristic of a microphone is its polar response. There are three basic types of polar response patterns: Omni-directional, Bi-directional, and Uni-directional (see figures 5 – 7 below). For use in a small studio to record a radio program, a uni-directional or cardioid microphone would best do the job. The reason for this is the fact that the cardioid mike has the greatest off-axis rejection of the three. This simply means it won't pick up stray noises such as cars passing by, motor noise of near-by equipment, or a squeaky desk chair.

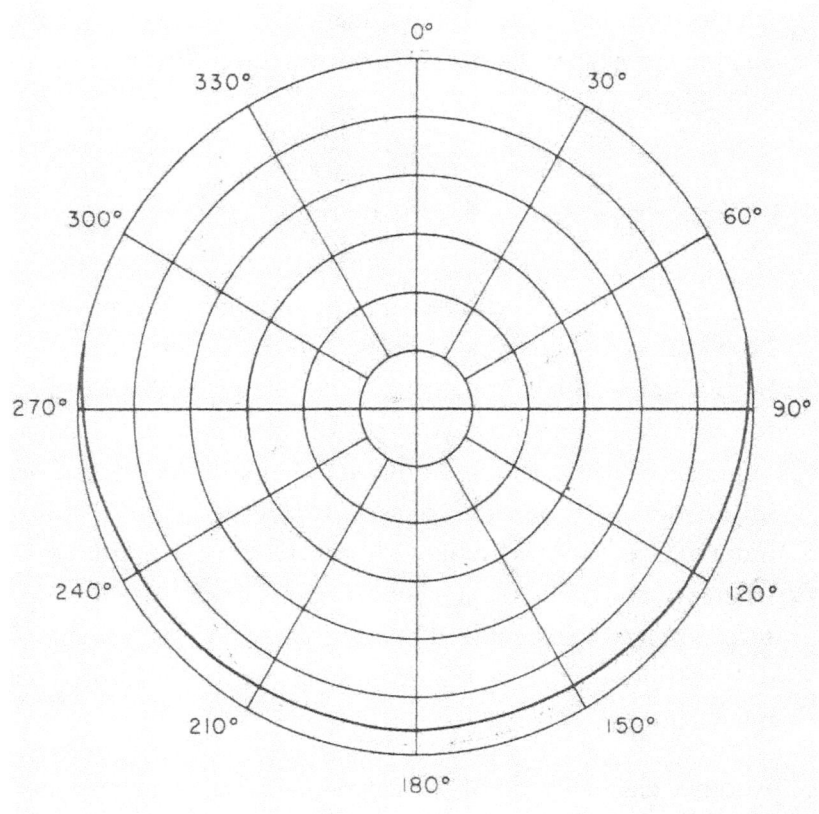

Omni-Directional Polar Pattern

An Omni-directional microphone will pick up about the same in any direction. The difference in front to back ratio is almost negligible (180 degrees off axis).

Figure 5

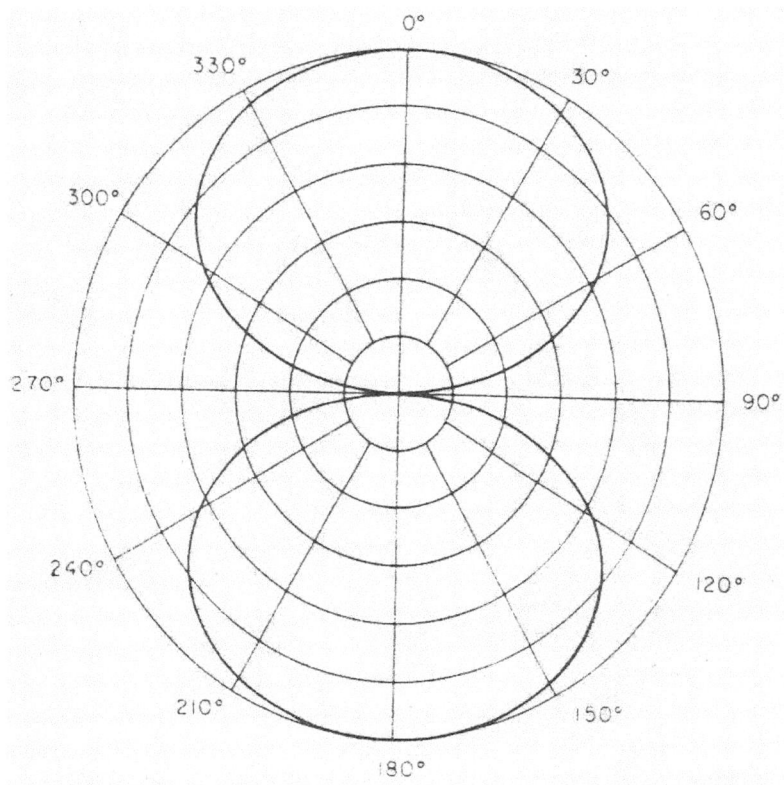

Bi-directional Polar Pattern

A Bi-directional microphone will have a symmetrical figure-eight shaped polar response with the region of maximum rejection at 90 degrees off axis, as shown in figure 6 above.

Figure 6

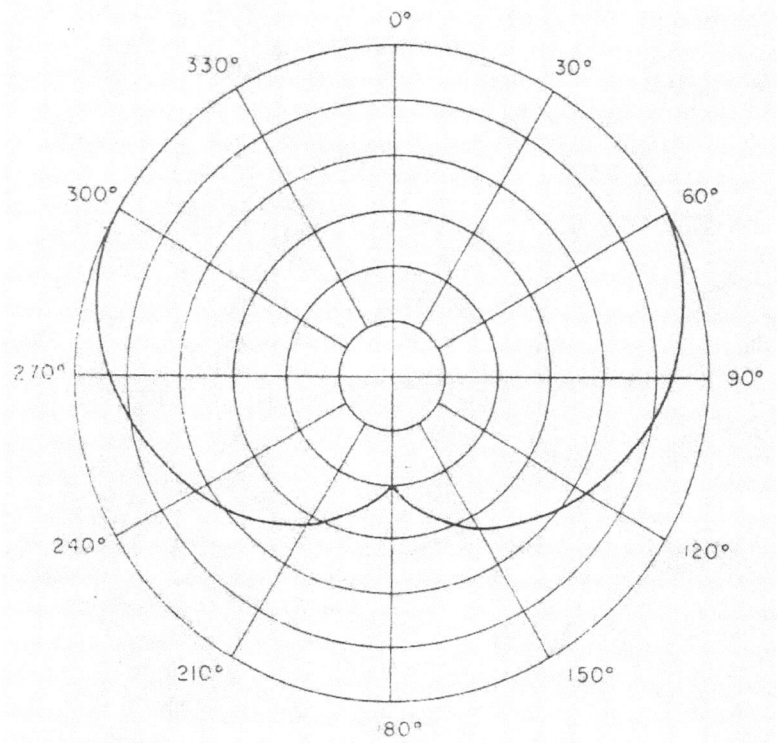

Uni-directional Polar Pattern

The Uni-directional or Cardioid microphone exhibits a polar response as shown above. The area of maximum rejection is at the rear of the mike (180 degrees).

Figure 7

If you are going to sit at a desk to do your program it is advised to mount the microphone on a floor mike stand with a boom so that the mike stand is not in contact with the desk. This will prevent unwanted noise from moving objects, books, or paper work during the program. You do not want to hold the mike in your hand while you record a program. Again the noise will be transferred to the recording along with the voice. The mike should be positioned ten to fourteen inches from the speaker's mouth depending upon the volume of the speaker. If you plan to preach (at higher volume than normal speech) into the mike you will want to back away to avoid over-loading. This will cause distortion unless properly processed with a limiter, compressor, or an automatic gain control (AGC). (I've never met a preacher that enjoyed preaching through a limiter.) You might also consider using a room that is not situated near a highway where there is an abundance of "road noise." This will be heard on your radio program unless the room is sound-proofed.

The monitor speaker should not be turned up while the microphone is on. The mike will pick it up and if loud enough begin to feedback or howl. When recording more than one person using a mike mixer, only those speaking should have their mikes turned up on the mixer or audio console. Do not use two microphones to pickup one voice. When you double the number of "open" microphones, you decrease overall gain by 3 dB. Additionally, there will be some phase cancellation as well as an audible phase effect heard on the air. Keep it simple and use a single microphone.

A pop filter or wind screen should be used for your studio mike. Normally this item is a Styrofoam or foam rubber cover over the face of the mike. Its purpose is to filter out the heavy "P" and "B" sounds from normal speech. These sounds are generally over-modulated. Many times wind screens are color-coded and are available in a variety of colors.

Audio Recording Devices

When this booklet was first published in 1983 we talked about reel to reel and cassette tape recorders in this section. Since that time many improvements have been made in Audio Recording Equipment.

While I would not recommend a cassette recording device for this purpose, there are a lot of quality reel to reel machines out there still doing the job. The old Ampex AG440 series were and still are excellent quality analog tape recorders. For many years these machines were the recording industry standard work-horse. Magnetic tape supplies may be an issue, but they also can still be found if that is the direction you want to go.

However, I recommend a digital audio recording device. There are several brands available in today's market place. I use a Tascam DR-40 portable digital recorder for my personal use. It has two built-in microphones and will record in one, two, three, or four separate tracks at a time with two extra microphone jacks. It uses an SD memory card to store the data and it is small enough to fit in a jacket pocket. This unit is small enough to place on a pulpit and record a sermon. You could use this same recorder in a studio and plug a mixer into it for use with not only a microphone, but playback devices for music, such as a CD player. You can tether this device to your laptop and transfer the material recorded (via WAV file) from the unit to the laptop. You can then email a copy of that file to a radio station or burn it into a CD and keep the file on your hard drive.

Similar digital recording devices are available by manufacturers such as Marantz, Olympus, Sony, Edirol, Zoom, Yamaha, Philips, et al. Most have built-in microphones. You will also see these units on television when several reporters are attempting to get a sound bite or interview with someone that is getting headlines. You'll see several hands with these devices held in front of someone's face recording what they say.

I used mine before and after church to get brief interviews with church members for several weeks. I built up a collection of short testimonies from members. In typical announcer style, I approached someone and asked their name. Then I asked them what it was about our church that they loved. Some took thirty seconds and some took a few minutes, but all of them had something good to say. When I had a total of about thirty sound bites, I burned them into a CD and gave it to the pastor. Every week on the thirty minute weekly radio program he started off the program with one of those recordings to encourage new people to come and visit. It was an excellent radio promotion.

Mixers and Audio Consoles

If you are putting together a five, fifteen, or thirty minute broadcast, one way to keep the cost down and keep it simple at the same time is to leave off the music. A fifteen or thirty minute program is long enough that it could be broken up into segments. One of the best ways to segment the program is the use of music. This helps the flow of the program, but at the same time requires much more equipment. For a voice-alone broadcast a recording device and a microphone are the only equipment necessary. If you decide to add the element of music, an audio mixer will be required.

The mixer or audio console is the main piece of equipment in a studio. Every piece of equipment is connected to this unit. The mixer is used to control the volume and blend program elements such as mike, recorded sound bites, and music resources. When connecting high-level inputs into a mixer, care should be taken to make certain that audio levels and impedance matching is correct. A qualified sound technician should be able to make the proper connections.

There are many small audio mixers on the market today that lend themselves perfectly to this application. These mixers will have microphone level and line level inputs. They will have line level outputs to feed a phone line or audio recording devices or both. They can be installed in the most simple manner or can be hooked up with all the whistles and bells. In the back of this booklet there will be a list of resources for audio equipment including audio consoles and mixers.

Labeling

It is always important that you make certain your program is labeled properly. The name of the program, the name of the preacher, and the date it is to be aired needs to be plainly labeled on the CD and CD Case, WAV file, or however your program is transported. Be sure that your program is easy to find, easy to spot, and easy to air. Radio stations accumulate tons of audio recordings. They get stacked up everywhere. One never knows when one's radio program gets misplaced somewhere.

Mailing

If you will be mailing your program to one or more radio stations, the same procedure should be followed concerning labeling. Make sure that your program can be identified from the exterior of the mailing envelope or box. Make certain that your address is plainly in view. Check your local post office for book rate and 4th class rates. You can save money by mailing early and not requiring first class postage on your programs.

Email Your Program

A great way to utilize modern technology is the use of email to send a radio program to the radio station. In order to take advantage of this method you will need a digital recording unit like the one mentioned earlier in this chapter. Record your program in your office or studio and attach the .wav file from the recording device to your email addressed to the radio station. This will entail tethering the device to your computer to make the transfer. This is a simple process and saves time and expense in the logistical side of a radio ministry. Conceivably, you could stay current with a daily broadcast from anywhere in the world using this method of delivery as long as Internet is available.

Audio Processing

It is recommended that you not be concerned with processing your audio in the production of your programs. This is an added expense and one that is not necessarily needed. The radio station on which you will be airing your program will have all the audio processing equipment needed. I have heard some programs which were improperly processed. This in addition to the station's processing made the program sound quite unnatural and "over-done." If your audio recorder has built-in AGC or limiter, switch it to the off position. This will assure a much more natural sound for your radio broadcast.

Miscellaneous

In order to make your office/study or radio room as conducive as possible to a professional sounding radio program, let me recommend a few additional things that factor into the big picture.

I recommend a good timing device in clear view during your program, especially if you are on live. The traditional black and white "Broadcast Clock" is an excellent choice. You have the correct time in front of your face all the time you are on the air, just like the announcer at the radio station. It is also a good idea to make sure you and the radio station show the same exact time. One of the best ways to get on the bad side of radio station personnel is to go over your allotted time. Keep in mind he has a program log in front of him and he MUST stay on time all day. There are no extra minutes in any of his hours. They are all accounted for. There are advertisements and station IDs and other things that must be aired in addition to your program. He will be your best friend if you are conscientious about going on and off the air. The preacher needs to be consistent and prompt every time he goes on the air. You simply cannot improve on common courtesy. It makes the announcer's job easier and he will appreciate your professionalism. Additionally, when you need a little favor from the radio people (and eventually you will), he will be there for you.

Another item I recommend for any radio studio is a large copy stand. In every radio station that I engineered, I was also one of the on-air announcers. In each of these stations I built a wood and plexiglass copy stand that sat on top of the audio console. I made them the same width as the audio console so there was plenty of room for news, weather, public service announcements, etc. This helpful device will provide plenty of room to place all your announcements. They can be easily read without the sound of rustling paper going out over the air. This will also eliminate searching for something that needs to be announced, since everything is right in front of your face. In many cases the computer monitor has replaced the copy stand, but not completely. I still recommend them for a radio ministry.

The preacher also needs to be aware of telephones ringing during the program. It goes without saying that the cell phone needs to be muted and the desk phone's ringer volume turned down as well. These are little things that contribute to a polished and professional sounding broadcast. Your listeners deserve the best and so does the Lord.

Depending on where it is that you do your recording or live broadcast, you might even consider putting an "On-the-Air" light outside your door. This will serve as a reminder for office personnel or visitors that when the light is on, it is not a good time to come calling.

It is also a good idea (when time permits) to record standby radio programs and have them handy. You never know what the future holds and it is wise to ready for the unexpected. You might come down with a cold or sore throat and your voice is simply not working properly. You might get called out of town. You might not have a choice and must be somewhere else during your broadcast hour. These spare programs can be in the possession of the radio station or a trusted worker or staff member at your church. They should know what to do when the time comes.

CHAPTER 6

Phone Lines

A phone line is definitely considered to be a part of your equipment. I have reserved an entire chapter to discuss this vital link between you and your radio audience. For the preacher who is considering a daily radio program and must make every minute and dollar count, then he should seriously consider doing his program over a phone line.

There are two basic methods of phone line use for on-air broadcasting:

The first is the standard telephone hook-up. This method simply utilizes the phone in your office or study as a link between you and the radio station. You simply dial the radio station's number before you go on and the announcer will patch you into the console in the control room. At the proper time he starts your intro (while you are listening to the station in your study), then on cue, you come on and begin the program.

There is available to you, through your local telephone company, a special type of telephone. This phone has a phone jack in the side and a switch on the front. This is the type of phone you want if you decide to use this method. The output of the mike mixer is plugged into the jack on the side of the phone. After you dial the radio station and make contact with the announcer, then you turn the switch on the phone. This turns off the mouth-piece of your telephone and connects the output of your mike mixer to the phone line. This enables you to use your office phone for both radio and normal use.

The second method is to have a permanent line installed by the phone company. This special line is a one-way circuit which links your microphone directly to the audio console or patch bay at the radio station. You do not have to place a call to the station. The only equipment you will need is a microphone and a mike preamp or mixer. The output of the mixer is connected to the phone line. The other end of the line is installed into the radio station equipment.

These special circuits are available in different frequency ranges. The standard "loop" is a 4 KHz line and has the same audio quality as a regular phone line. Depending on your local telephone office, you can get 4, 6, 8, 10, or 15 KHz equalized lines installed. Anything over 4 KHz will be adequate. However, if yours is a special radio ministry requiring a lot of music, then you would be better off with an equalized line to provide a quality on-air sound.

If you want all the quality you can get and you are on an AM station, you would not need any more than a 7 KHz loop. The reason is that the AM broadcast on-air frequency response is less than that of FM broadcast. If you are on an FM station, then we recommend at 15 KHz equalized line in order to maintain the highest quality music reproduction. A phone call to your local phone office will give you an idea of the monthly costs of such lines.

The average daily broadcast over the phone should utilize a standard 4 KHz line. This line will pay for itself in a short period of time when compared to driving to the radio station on a regular basis.

The following is a typical scenario for a daily, local economic broadcast that has worked for many preachers and could work for you:

Two or three minutes before the program's air time the preacher calls the station on the regular office telephone. He checks in with the announcer and chooses a particular song for the announcer to cue up. The preacher switches the line over to the output of his mixer or mike preamp and hangs up the phone. Now his mixer is connected to the radio station. Note: The monitor (radio) should be turned down enough that it will not feed back during the broadcast. It might work better if the preacher uses a set of headphones to monitor to assure no feedback. After his intro is played, the preacher greets his audience, makes some announcements, then says, "Let's listen to this song…" At this point the announcer at the radio station plays the recording and puts the preacher's loop in cue until the record is over, then puts him back on the air. The preacher continues with his program until finished; then the announcer plays the outro; the broadcast is finished and the preacher never left his study. He has given his message over a 4 KHz loop and yet still had quality sound for his music portion since was played at the radio station and not over the phone line. This will work for five, fifteen, or thirty minute daily or weekly radio programs.

If you are not able to be in your study (at the mike) at your regular program time, it is possible to request the radio station personnel to record your program at an earlier time to be played back at normal time. This still utilizes your special phone circuit in your study or office while the station uses their recording equipment. Your personal cell phone also comes in handy with this same scenario. If you are away from your study and even out of town, you have the technology to do your broadcast completely by cell phone.

The folks at the radio station would probably appreciate a couple of spare programs on CD for emergency use. Sometimes Murphy's Law takes over and it is good to have at the station some pre-recorded programs ready to fill in just in case.

GLOSSARY

60 Cycle Hum - A low frequency noise commonly heard in PA systems, which can result for various reasons.

70 Volt Speaker System – A type of PA system using a number of parallel-wired, transformer/speakers which works well for large areas and long lengths of cable.

AC – Alternating Current

Acoustic Absorption – The ability of a material to dampen sound waves and contribute to the elimination of echo.

Active Component – Any device in the audio chain that requires power to operate.

Adjacencies – Programs on the air before and after a particular time slot.

Afternoon Drive – One of the two most heavily listened to times on a radio station: during the afternoon hours from 4 to 7 PM.

AGC – Automatic Gain Control - A method of maintaining a constant output level on an audio circuit.

Ambient Noise – The background noise in an environment.

Amplifier – An electronic component used to increase the amplitude of audio, voltage, or current.

Amplitude Modulation (AM) – A method of modulating a radio frequency carrier by audio modulation of the carrier amplitude.

Attenuator – A resistive device to reduce a signal level and maintain impedance matching.

Audio Frequency Spectrum – The frequency spectrum from roughly 20 to 20,000 Hz. based on human hearing range.

Audio Processing – To change the natural characteristics of an audio signal with use of AGC, limiters, equalizers, etc.

Audio Transformer – A passive device used in audio for impedance matching, isolation, RF elimination, and other uses.

Balanced Line – An audio cable consisting of a pair of wires surrounded with a shield used to ground out induced voltages and RF.

Band-pass Filter – An active or passive network of resistance, capacitance, or inductance, placed in the audio chain to allow a particular band of frequencies to pass through.

Bandwidth – A measurement in Hertz for the analog width of a spectrum of frequencies. For digital components it is the rate of data flow.

Bass Trap – An acoustic panel placed in an audio environment to offer a dampening effect to lower audio frequencies.

Bi-directional Microphone – A microphone which has a symmetrical firure-8 shaped polar response with the region of maximum attenuation at 90 degrees off axis.

Broadcast Day – A period of time between local sunrise and 12 midnight local time.

Cannon Connector – A common audio cable connector for balanced low impedance lines; also referred to as XLR connector.

Capacitor – A passive component used in numerous electronic functions. It has the ability to store electrical energy and dissipate the energy when needed. It is made up of two conductors separated by a dielectric insulating material.

Capacitive Reactance – Opposition to the flow of current in an AC circuit due to the properties of capacitance in the circuit.

Cardioid Microphone – A uni-directional microphone which has a polar response curve like a cardioid. The maximum attenuation is at the rear and gradually decreases toward the axis point.

Carrier Frequency – A certain frequency assigned to a station on which its program is modulated.

Clear Channel Station – A dominant AM station which operates 24 hours a day at 50,000 watts.

Clipping – The result of an audio signal when it is overdriven or driven past the power amplifier capabilities. The result is distortion.

Compression Amplifier – An active component in an audio chain where a relatively constant output level is needed. The amplifier reacts with attenuation to an extreme increase in signal level to avoid clipping.

Condenser Microphone – A microphone that uses a DC voltage across a capacitor to produce an audio signal. One of the plates of the capacitor is a diaphragm. As acoustic energy moves the diaphragm the voltage is modulated, thus producing the output signal. The mike requires an outside or phantom source of DC voltage.

Crossover Network – A network of inductors, capacitors, and resistors used to divide an audio signal into upper and lower frequencies, normally found in speaker cabinets. This network divides frequencies into lower, midrange, and high frequencies for appropriate transducers.

Cuable Pot - A volume control on an audio console which as a special cue channel to audition the program element or cue up audio.

Current – The flow of electrons in an electric circuit. It can be compared to water flowing through a garden hose, where the hose represents the circuit and the force of the spigot (drive) is voltage which pushes the current through the circuit.

DA – Distribution Amp – provides multiple outputs of a single audio source.

Daily – A radio broadcast which is aired Monday through Friday.

Day Part – The division of a broadcast day.

dB - Decibel - A logarithmic unit used to define the intensity of an audio signal. An increase of 3 dB is equal to doubling the power.

DC – Direct Current; as opposed to Alternating Current, DC is a continuous voltage.

Decay Time – AKA T60 - With respect to echo and reverberation, the time it takes for audio to decay by 60 dB.

De-esser – A device that attenuates or eliminates sibilant sounds in recording or reproducing the human voice.

Directional Antenna System – The use of two or more towers at an AM broadcast station to direct the signal in a given direction and inhibit the signal from radiating in another direction.

Distortion – An unwanted or wanted (depending on the application) change in an audio signal.

Doppler Effect – This is an audio phenomena that is heard when an audio source is in motion relative to the listener. To use a common experience, let's say you hear the sound of an aircraft approaching. The sound of the engine will increase in pitch (frequency) as it approaches. Then as it passes overhead and moves away, the pitch reverses and goes down. While the pitch of the engine never changes, it sounds like it changes because of the speed of the aircraft relative to the speed of sound at your point of listening.

Dummy Load – A resistive component that absorbs all the output power of a transmitter or amplifier to simulate working conditions for purposes of testing.

Dynamic Microphone – A device to transfer the acoustic energy of sound into electrical energy by utilizing a diaphragm and magnetic induction.

Electret Microphone – A relatively inexpensive type of condenser microphone that uses battery power.

Equalized Line – A telephone loop between two points which maintains the same audio quality at the output as the input.

Equalizer – Based on the root word, equal, an equalizer is an audio device whose function is to equal out the tonal characteristics of a sound. At least that was the idea back in the days when they were first conceived as a tool used to get flat response in telephone lines and to make up for the deficiencies in audio equipment and acoustic spaces. Nowadays it could more aptly be named an "unequalizer" since they are more often used creatively to alter the relative balance of frequencies to produce desired tonal characteristics in sounds. An equalizer has the ability to boost and/or cut the energy (amplitude) in specified frequency ranges by employing one or more filter circuits. There are many different types of EQ's in use today in many widely varying applications, but they fundamentally all do the same thing. (Sweetwater.com)

ERP – Effective Radiated Power: The total amount of power being broadcast or the sum of the transmitter output and antenna gain. For example, you might have an FM broadcast transmitter that puts out 18 Kilowatts that feeds an antenna system that multiplies that power by the number of bays that are stacked on the tower. The ERP could easily be 100 Kilowatts.

Fader – Another name for variable attenuator, volume control, or potentiometer. A fader works just like a standard potentiometer, only instead of rotating, it slides along a straight path. Faders are most commonly used on mixing boards and graphic equalizers, where it is nice to have both an easy way to move the level up and down, and to provide a sort of graphic representation of the relative levels of many channels (or frequencies in the case of an EQ). Faders have also historically been used on some synthesizers as controllers for various parameters. The name comes from the phrase "fade out." Once faders came into existence it became much easier for an engineer to do a smooth fade out. (Sweetwater.com)

Feedback – The howling sound or oscillation present when the output of an amplifier is feeding the input and a closed loop is completed.

Flat – A frequency response which is generally the same over a given band width, which for audio would be 20 to 20,000 Hz.

Fold back – Another term for a monitor system for the stage area.

Frequency – The number of times an alternating current goes through its complete cycle per second.

Frequency Modulation (FM) – A method of modulating a radio frequency carrier by alternating the frequency of the carrier.

Frequency Response – A rating of how effectively a circuit or device transmits the different frequencies applied to it.

Gain – Volume

Ground Loop – In an audio system, a ground loop can result when there are multiple components (active and passive) wired together that ideally are the same electrical potential. With more than one power supply it is possible to have differing voltage potentials between these components. This can result if there exists more than one ground point for multiple grounds. For example a piece of equipment may be grounded to the rack to which it is mounted. Another component may be grounded to another source that isn't connected electrically to the rack. There could be a potential difference of 5 volts between the two grounds, which could produce a hum into the system. More severe differences could be caused by incorrect wiring and lead to voltage differences of up to 50 volts or more. These differences in ground or voltage potentials, in many cases are the cause for hum in a system. In the early days, before a ground pin on electrical plugs was standard, you could play an electric guitar and get shocked if your mouth touched the microphone on the PA system (I know how this feels). The use of a common ground and correct electrical wiring can eliminate these issues.

Harmonic – A frequency that is an integral multiple of a fundamental frequency.

Hertz – (**Hz.**) Unit of measure for frequency (Cycles per second).

Impedance Matching Transformer – A transformer used to match two different impedances; for example, a high impedance unbalanced line to a low impedance balanced line.

High Input or Output Impedance – The rated impedance across the input or output of an audio component ranging from 10,000 to 100,000 ohms.

Impedance – The total amount of resistance to the flow of current in an AC circuit, including DC resistance, inductance reactance, and capacitance reactance.

Impedance Matching – This is the correct practice for connecting audio components in a sound system and will result in the maximum transfer of signal and audio fidelity. For example: the 8 ohm output of a headset jack is not a good source to feed the high impedance (50K ohm) input of a recording device. In this case the use of a transformer to match the impedances would result in a much better sounding signal.

Inductance – The electromagnetic effect produced by current flowing through a coil of wire, which has the tendency to oppose any change of current.

Inductive Reactance – Opposition to the flow of current in an AC circuit due to the properties of inductance in the circuit.

Inductor – A coil of wire which, when current flows through it, generates an electromotive force which tends to oppose a change of current.

Intelligibility – The ability to be clearly understood.

Intermodulation Distortion – the undesired interaction of electronic signals of different frequencies transmitted within a nonlinear system, resulting in distortion.

Intro – A brief recorded or live announcement at the beginning of a program giving the name of the program, speaker, etc.

Kilohertz – (**KHz**) Unit of frequency equivalent to 1000 Hz.

Kilowatt – (**KW**) 1000 watts

Limiter – An amplifier circuit in which the amplitude of the output is prevented from exceeding a given value. Thus, it can be used to remove any amplitude variations in a signal while leaving an FM signal intact.

Line Level – A power level for an audio signal from approximately -10 dB to +20 dB into 600 ohms.

Line Loss – A gradual decrease in the quality and strength of a signal as the length of the cable increases.

Loop – A term used for a special phone line for broadcast use.

Low Impedance – 8 to 600 ohms

Masking Noise – Often found in large offices with multiple personnel; the use of sound (for example pink noise) to generate an ambient noise level for the purpose of covering up phone conversations and masking other audible information.

Microphone Boom – A mike stand used to extend the mike horizontally or diagonally.

Microphone Level Input – Approximately -50 to -60 dB (much lower than line level)

Microphone Mixer – A unit consisting of mike preamps, switches, and potentiometers (pots), used to blend and switch mikes and other program elements.

Microphone Preamp – An amplifier used to boost mike output (-60 dB) to line level which is from approximately -10 to +20 dB.

Middle C – The note "C" that is found approximately in the middle of a full sized piano. It is commonly tuned to 261.1 Hz. The pitch is represented in written music as the note on the 1st ledger line below the treble staff, or the note on the 1st ledger line above the bass clef. (Sweetwater.com)

Monaural – A single channel system, as opposed to stereo or two channel system. Also known as mono.

Monitor Speaker – An audio transducer used in a control room which monitors program material.

Morning Drive – A heavily listened-to time of a broadcast day between 6 and 10 AM.

Noise – Any sound or combination of sounds that is unwanted in an audio system.

Notch Filter – An electronic network, passive or active, utilizing resistance, capacitance, and/or inductance that tends to reduce the amplitude of a particular frequency; often used to eliminate feedback in conjunction with a room equalizer.

Off Axis Rejection – A term used regarding mike response; having to do with the amount of attenuation or noise rejection that a microphone has on both sides and behind (off axis).

Ohm – Unit of measure for resistance and impedance named after German physicist Georg Simon Ohm.

Omni-directional mike – A microphone which picks up the same amount of sound from all sides.

Omni-directional antenna – Similar to an omni-directional microphone, a radio station can have an omni-directional signal, which radiates equally in all directions.

Oscillator – An amplifier with a feedback circuit that induces oscillation at a particular frequency.

Outro – A brief recorded or live announcement at the close of a program giving the name of the program, speaker, etc.

Pad – A device inserted into a circuit to introduce transmission loss or attenuation and maintain impedance matching.

Pan – To adjust an audio signal between left, center, and right, as well as front to rear.

Parallel – A method of connecting electronic devices or components for operation, so that all of the hot leads are connected and all of the low sides are connected; as opposed to "Series."

Passive – A device that does not require power to operate.

Patch Panel – Also referred to as a patch bay or jackfield, a panel that houses multiple jacks or cable connections. Patch panels can be used for audio, video, telephone connections, and data transfers.

Peak Limiter – An active audio component that limits the output of an audio signal to prevent overloading the input of the subsequent audio component in the chain.

Phantom Power Supply – A supplier of a DC voltage used to power condenser microphones. This can be anywhere from 5 to 48 volts. In many cases this power supply is built in to the audio console.

Phase – Important to remember regarding wiring pairs of audio cables; Problems can arise if a pair is out of phase with the others.

Phase Cancellation – Phase describes where, in its cycle, a periodic waveform is at any given time. The relationship in time of two or more waveforms with the same or harmonically related periods gives us a measurement of their phase difference. Phase cancellation occurs when two signals of the same frequency are out of phase with each other resulting in a net reduction in the overall level of the combined signal. If two identical signals are 100% or 180 degrees out of phase, they will completely cancel one another if combined. When similar complex signals (such as the left and right channel of a stereo music program) are combined, phase cancellation will cause some frequencies to be cut, while others may end up boosted. Phase and phase difference is a real-world issue in areas such as electrical wiring of audio equipment, signal path, and microphone placement during the recording process. Phase reversal can be a serious compromise of sound quality or a special effect affecting the perceived spaciousness of the sound depending on the context of its occurrence (Sweetwater.com).

Phone Jack – A ¼" female receptacle into which a ¼" phone plug is inserted; used for unbalanced audio including headphones, speakers, patching, etc.

Pink Noise – An audio signal made up of random and spurious audio frequencies with a gradual increase in amplitude with decrease in frequency. The boost at the low end gives it a "roaring" sound similar to a waterfall or ocean waves rather than the "hiss" sound of white noise, which has a flat frequency response. Pink noise is commonly used to equalize a PA system.

Pitch – Another term for frequency. For example 440 Hz. is equivalent to middle "A" (pitch) on a piano.

Polarity – The proper orienting or arrangement of wire pairs with respect to hot side and ground or hot (high) and low plus ground.

Polar Response – Graphically, a diagram which shows the magnitude of a quantity in relation to direction. Or, the characteristics of a microphone, for example, having to do with front to back ratio, front to side, etc.

Pop Filter – (Wind screen) a filter placed over a microphone to remove the over-modulated "P" sound and "B" sound from normal speech; usually made from Styrofoam or cloth and many cases color-coded.

Pot – Short for potentiometer or variable resistor used as a volume control.

Pre-emphasis Kick – A gradual increase in amplitude with an increase in frequency (towards the upper end of the audio spectrum) for the purpose of improved signal to noise ratios and clarity. This concept has been used in the shaping of the frequency response of lavaliere microphones, broadcasting, recording, et al.

Pre-sunrise Authorization – A special authorization given to daytime AM radio stations which allows them to sign on before sunrise at a reduced power. Usually at 6 or 6:30 AM.

Presence – In movie-making and television production, presence (or room tone) is the "silence" recorded at a location or space with the absence of dialogue.

Promo or Promotional Announcement – An announcement made on the air promoting a program or event happening at another time.

Radio Market – A term used in broadcast for the business area a station serves.

Radio Format – Having to do with type of programming a radio station broadcasts; whether Gospel, country, etc.

RC Network – A passive array of resistors and capacitors used to pass, block, or change the amplitude of a particular frequency or band of frequencies.

Real Time Analyzer – (RTA) is a professional audio device that measures and displays the frequency spectrum of an audio signal; a spectrum analyzer that works in real time. An RTA can range from a small hand-held device to a rack-mounted hardware unit to software running on a laptop.

Resistance – The measure of the amount of opposition a device offers to the flow of current. The unit of measure is the ohm. Resistance in a DC circuit is measured with an ohm meter.

Resonance –This is the propensity of a system to oscillate at a specific frequency with a greater amplitude than other frequencies. In many cases there is a "family" of resonant frequencies (often a fundamental frequency and its harmonics) which turn out to be incremental when measured. The fundamental frequency will always carry the greater amplitude.

Reverberation – The remainder of sound that exists in a room after the source of the sound has stopped is called reverberation, sometimes mistakenly called echo (which is an entirely different sounding phenomenon). We've all heard it when doing something like clapping our hands (or bouncing a basketball) in a large enclosed space (like a gym). All rooms have some reverberation, even though we may not always notice it as such. The characteristics of the reverberation are a big part of the subjective quality of the sound of any room in which we are located. (Sweetwater.com)

RF – Radio Frequency –A band of frequencies which include part of the audio spectrum as its low end and a high frequency that merges with the infrared frequencies (9 KHz to 300 GHz). These frequencies can affect church PA systems and can be radiated by the following sources: AM and FM broadcast stations, CB radios, VHF and UHF TV stations, Ham radios, wireless microphones, garage door openers, baby monitors, air traffic control, alarm systems, cordless telephones, microwave ovens, et al.

RF Filter – Essentially, a low-pass filter to allow audio frequencies, but block everything above that.

RMS – Root Mean Square; true power reading.

Roll Off (or Rolloff) – Specifically roll off refers to the action of a specific type of filter; one designed to roll off frequencies above or below a certain point. It is called roll off because the process is gradual. Hi pass and low pass filters both roll off frequencies outside of their range, but they don't immediately eliminate all frequencies outside their range. The sound is gently (or not so gently) "rolled off" with frequencies further above or below the cutoff frequency becoming more and more attenuated. Roll off steepness is generally stated in dB per Octave, with higher numbers indicating a steeper filter. 24 dB/Octave is steeper than 12 dB/Octave. (Sweetwater.com)

RTA – see Real Time Analyzer.

Rumbling – Unwanted noise from motorized equipment, including motor noise, HVAC, et al, which finds its way into the audio circuit.

Series – A wiring protocol where devices are connected sequentially as opposed to in parallel with one another. Specifically this means that each device is an integral component in the circuit path, such that if one device fails the circuit will become open and no longer function. For example, a circuit breaker is typically wired in series with the hot wire going to each circuit in a house. When the breaker opens the circuit no longer has power. (Sweetwater.com)

Signal to Noise Ratio – The name defines itself. In an audio system the audio signal should be at a much higher level than the intrinsic noise level of the system. The two levels are compared with each other in the S/N ratio. The higher the signal level is to the noise level; the better the audio quality will be. For example: if you don't set a microphone volume control high enough and compensate by raising the master board output, you will decrease the signal to noise ratio and thus increase the noise level. The mike pot should be moving the VU indicator up to the 100% level to eliminate this problem.

Snake – A cable assembly used in audio containing multiple audio cables. Typically it would have up to 50 or more individual shielded pairs all contained in one protective jacket. Normally one would see the cables terminated with XLR connectors or quarter-inch phone plugs. A snake cable is normally used to connect microphones and monitoring equipment to an audio console located away from the audio source area.

Sounding Board – A rigid surface placed above and/or behind a pulpit or speaking platform for the purpose of projecting the voice of the speaker. An illustration of this is the top of a grand piano when it is propped up to deflect the sound out to the audience.

Spectrum Analyzer – An instrument which gives a graphic display of the audio spectrum. In audio PA systems it is used in conjunction with an injected signal such as pink noise to adjust the sound system for maximum gain before feedback as well as room equalization. The display will indicate the overall frequency response of the room being equalized.

Spike – An abrupt transient that forms part of a pulse, but is of greater amplitude than the average value of the pulse. It is often heard in the speaker as a pop or thump.

SPL – Sound Pressure Level

Spot Sales – Term used in commercial radio for sale of spot commercials played on the air during music programs.

Standard 4 KHz Phone Line – Standard bandwidth of telephone line.

Standard Broadcast Band – Between 535 KHz and 1605 KHz

Stereo – A prefix meaning three-dimensional or an audio system utilizing left and right channels.

STL – Studio Transmitter Link; either microwave, fiber, IP, or phone line.

Sub-audible Frequency – A frequency too low to be heard by the human ear (20-20K Hz.).

Syndicated – Refers to a radio program distributed to a vast number of radio stations simultaneously.

Talkback Mike – A microphone at the audio console for the purpose of the sound tech speaking to people in a recording studio or sound stage area.

Transducer – A device that transforms electrical energy into acoustic energy or vice versa.

Tweeter – An audio loudspeaker designed to reproduce high audio frequencies in the approximate range of 2000 Hz to 20,000 Hz.

Unbalanced Line – An audio line consisting of a single conductor with a shield. This type of line is used in consumer audio equipment utilizing RCA, phone plug, and mini-plug connectors.

Uni-directional Microphone – A cardioid microphone which exhibits a polar response in the shape of a heart. It has its highest noise rejection at the rear of the mike.

Voice Frequency –This is a frequency range that consists of those frequencies which are generally used by the human voice. Specifically, this band is from approximately 300 Hz to 3400 Hz.

Voltage – The electromotive force that moves current through an electrical conductor. Voltage produces either DC or AC.

VU Meter – Abbreviation for Volume Unit meter. A voltmeter used to measure audio voltage.

Wave File (.wav) – This is a file format used for audio applications in computers and audio recording devices.

Wave Length – The distance between points of corresponding phase of two consecutive cycles.

Wind Screen – AKA Pop Filter – Used over a microphone to filter out excessive over modulation of "P" and "B" sounds in speech.

White Noise – An audio signal made up of random and spurious audio frequencies with a flat frequency response. You can sample white noise by listening to the "hiss" sound when tuning an FM radio between channels.

Woofer – A loudspeaker driver designed to produce low frequency sounds, typically from around 40 hertz up to about a kilohertz or higher.

XLR – See Cannon connector.

Audio Equipment Resources:

Listed below are a few websites of businesses that can supply you with any and all audio equipment needs. There are more out there, but these will get you started.

www.musiciansfriend,com
www.americanmusical.com
www.music123.com
www.zzounds.com
www.guitarcenter.com
www.bswusa.com
www.proacousticsusa.com
www.sweetwater.com

A WORD ABOUT THE AUTHOR

Charlie Edwards worked for a number of radio stations from Kentucky to Tennessee to Florida. He has worn most of the hats in a radio station: Announcer, Chief Engineer, Program Director, Sales Rep, and General Manager. He also has designed and installed sound systems in churches as well as built radio station control rooms and production rooms.

He completed an Electronic Communications training program and graduated at the top of his class as well as earned an FCC First Class Radio Telephone Operator License in the early 70s. He is a member of the Society of Broadcast Engineers and holds five certifications including, Certified Audio Engineer, Certified Broadcast Radio Engineer, Certified Broadcast Television Engineer. He graduated with a BA from Tennessee Temple University. He also earned a Master of Theology, Doctor of Theology, and most recently he finished up the requirements for a PhD in Clinical Christian Counseling.

Dr. Edwards wrote a novel entitled, "Caribbean Sentinel." He also wrote a book for Pastors, Worship Leaders, and laymen sound technicians, entitled, "Church PA System Handbook." He has also published a collection of Theological essays entitled, "Bible White Papers." All of these works are available on Amazon.com in both paperback and download format.

It is our hope that this booklet will be a help to pastors and laymen who desire to expand their outreach ministry in the fruitful area of radio broadcasting.

www.ingramcontent.com/pod-product-compliance
Lightning Source LLC
Chambersburg PA
CBHW071415040426
42444CB00009B/2258